CYBER SAFETY

QUESTIONS & ANSWERS BOOK

Written by Molly Sanchez | Designed by Anna Asfour

QUESTION...

What is identity theft, and how can I prevent it from happening to me?

ANSWER:

Identity theft is when people steal your personal information—usually using your social security number—to get loans, credit cards, or even medical services. The problem with this type of fraud is that it could destroy your credit. This means that when you are 18 or older and you apply for a loan to get a car or home, creditors will look at your credit report. If your information has been used fraudulently, they may find that you have a terrible credit score because the person using your information has not paid his or her bills on time, or at all.

CONTINUED...

That will make it so lenders will not trust you and could make it very difficult for you to get a loan, or you will be charged very high interest rates, costing you a lot of money. Additionally, it can be costly and take a lot of time to repair your credit rating.

To prevent this from happening to you, do not fill out online forms with your social security number. Avoid giving out other personal information online, especially on public internet servers and nonsecure websites, like in a hotel or an airport. You should see a padlock next to the web address for secure websites. Another important thing you can do to prevent identity theft is create strong passwords to prevent hackers from accessing your information.

QUESTION...

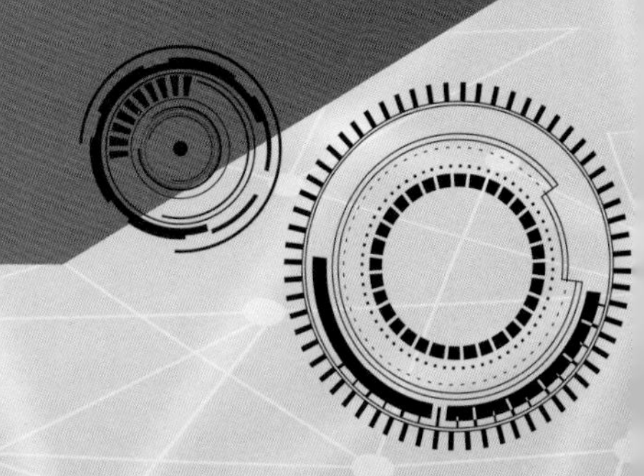

How can I create strong passwords?

HERE ARE SIX RULES TO FOLLOW WHEN CREATING A PASSWORD:

1. Make it long. The longer it is, the harder it will be to guess or hack.
2. Don't use personal information. Avoid using names, birthdays, or your house or street numbers.
3. Include uppercase and lowercase letters, as well as numbers and symbols. Remember, you are trying to make it impossible to guess.

CONTINUED...

4. Don't use words that can be found in a dictionary. They make weak passwords.

5. Don't use the same password for multiple accounts. If you do and one account is hacked, your other accounts become easy targets.

6. Create long, random passwords. For instance, think of a phrase, such as "My account is safe from hackers." Use the first letter of each word to create a password, capitalizing some of them: mAisfH. Then add numbers and symbols: m$Ai#sf?H. You can use a secure password manager to remember online passwords for you.

QUESTION...

Is it okay to copy and paste things that I find online?

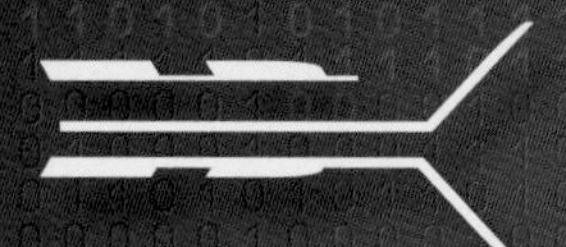

ANSWER:

This is only acceptable if you give credit to the person who created the content. If you quote someone, acknowledge the quote is not yours by attaching the author's name to the quote. If it's a photograph, give the photographer the credit. Provide a link where appropriate. It is not okay to copy and paste sentences and paragraphs and use them as your own.

QUESTION...

What are online predators?

Online predators are people who use the internet to find teens and kids they can trick into sexual behavior or abuse in other ways.

The FBI says that there are 500,000 active predators on the internet every day with multiple profiles. Their main target is kids between 12 and 15.

CONTINUED...

It is very important to tell an adult if someone online offers you money, gifts, or anything else in exchange for photos or videos of you.

If you have sent photos or videos, understand that this situation is NOT YOUR FAULT, and you will not be in trouble if you tell an adult. The predator has committed a crime, not you.

QUESTION...

What is sextortion?

ANSWER:

This is a crime committed by online predators. They pretend to be someone young and attractive with stolen photos and fake information about themselves. Their goal is to lure in young people and get them to send nude photos and videos. Then they will use those photos or videos to get more by threatening to send them to your friends or family members.

Many youth are frightened and shamed into submitting to predators' demands for more and more nude and sexual content. If this ever happens to you or someone you know, it is vitally important to tell an adult. You will not be in trouble. You have been a victim of a crime, and adults want to help put an end to such abuse and crime.

QUESTION...

How do I avoid becoming a victim of online predators?

Avoid talking to people online that you don't know in real life. Recognize if anyone online ever asks you for anything inappropriate, and end the conversation immediately. Tell an adult if you ever feel uncomfortable about an online conversation.

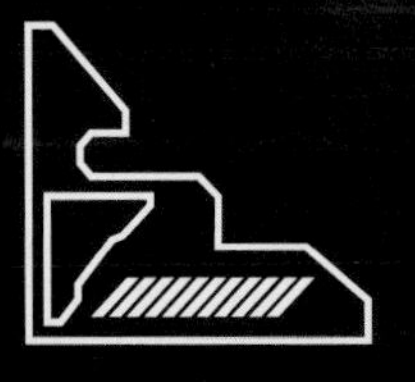

QUESTION...

What is phishing, and how can I recognize and avoid it?

Phishing is the act of sending fake emails (or other communication) that appear to be from a reliable source, such as your bank or a police department. The goal is to get you to share personal information or to click on a link that may download malware onto your device.

CONTINUED...

Pay close attention to the email address. Is it one you recognize? If you are unsure, check with a parent or other trusted adult.

Be careful not to click on links unless you are certain they are from reputable companies.

Never give out your personal information online unless you know you are on the company's real website. The best defense against phishing is learning about it.

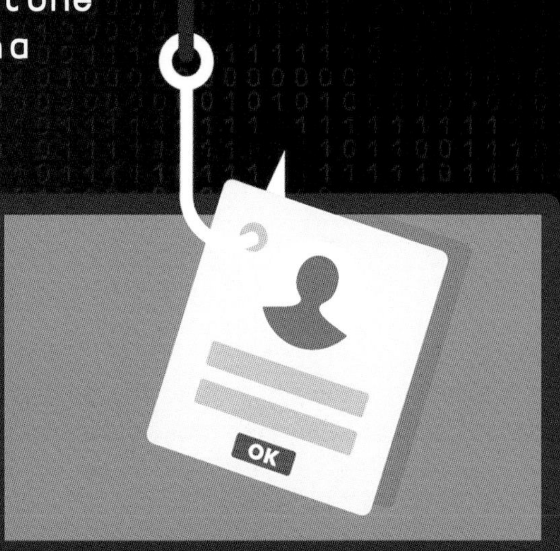

QUESTION...

What is malware?

ANSWER:

Malware is a nickname for "malicious software." It is created to cause damage to, and possibly destroy, computers and computer systems. Some types of malware can steal information or lead you to harmful websites. Other types will send you an overload of ads (called spam). Still others can monitor what you do on your computer, even accessing your webcam to watch you.

QUESTION...

How can I avoid malware and protect my computer from viruses?

A good antivirus software is the best way to avoid malware and viruses. It can detect viruses and malware if they have made it into your computer. Also, be very careful about what links you click on and things you download.

QUESTION...

How can I avoid bait and switch tactics?

Bait and switch is when a company lures you in with a really great "deal," but then it turns out the item advertised is not actually available. Then they try to sell you a more expensive item or a much lower-quality item for the advertised price.

1. If an offer seems too good to be true, it most likely is.
2. Do some research on the company before you buy.
3. Avoid quick impulse buying.

QUESTION...

What if I am asked for my personal information online?

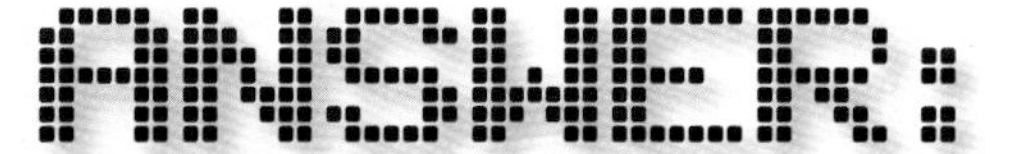

Companies and individuals should NEVER ask for the personal information of a minor. If they do, tell a parent or another adult and don't give out any information.

QUESTION...

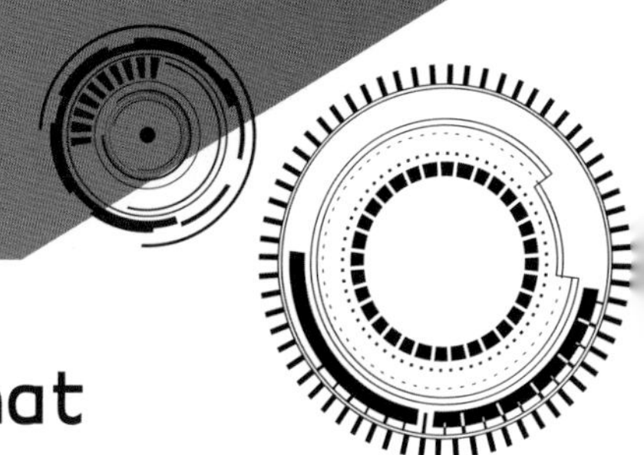

What should I do if I see something online that makes me uncomfortable?

ANSWER:

Tell a trusted adult. Remember, you are not in trouble. It is always better to talk to someone about something that is bothering you than to carry a burden quietly on your own. Your parents and others want you to have a safe online experience. Despite the dangers, there is so much good that can come from using the internet.

QUESTION...

How can social media be harmful?

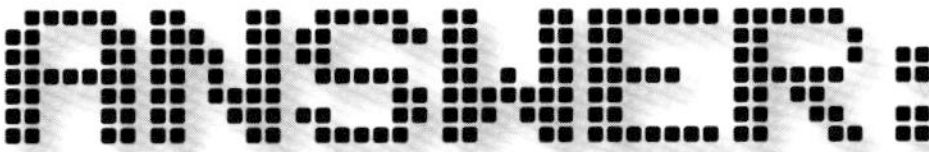

- Studies show an increase of depression among teens and adults who use social media regularly, as well as a rise in anxiety and a decrease in self-esteem.

- Comparison can be harmful and dangerous, as we compare our lives with the seemingly perfect lives of everyone else.

- People who spend too much time on social media and texting miss out on real, face-to-face relationships and the social skills that come with them.

- A lot of bullying happens on social media; people will say things behind the safety of a screen that they would never say face-to-face, and the results can be devastating.

QUESTION...

Why is kindness important in my online interactions?

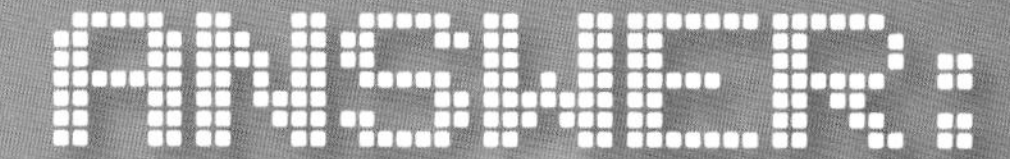

There is plenty of negativity online. People can get pretty mean when they are not facing anyone in person. As the saying goes, a little kindness goes a long way. You never know the positive impact you could have by paying someone a compliment, leaving a positive comment on a post, or sharing an inspirational message.

CONTINUED...

Consider how rampant anxiety and depression are; you could really make a difference for someone who needs a lift. The Golden Rule is a good guide: don't say anything you wouldn't want someone to say to you. Kindness is also the best tool to defuse a heated situation.

QUESTION...

What are online addictions, and how can I avoid getting sucked in?

ANSWER:

When any online activity becomes a hindrance and interferes with a person's normal daily activities—like homework, work, or sleep—it can be called an addiction. Online addiction can include online relationships, online shopping, compulsive information seeking, online gaming, excessive time on social media, online gambling, and pornography.

If people frequently tell you that you spend too much time online, you may have an addiction. Talk about it with a trusted friend or adult. Set parameters for yourself and share them with someone else for accountability. If needed, get professional help.

QUESTION...

Why is pornography harmful?

ANSWER:

Pornography is harmful to everyone. There is a long list of ways that pornography can harm people physically, socially, mentally, and spiritually. Some of these effects include depression, inability to connect with others, sexual dysfunction, addiction, and a disregard for one's identity as a child of God.

QUESTION...

What should I do if I see pornography or other inappropriate content?

ANSWER:

Whether seeing pornography or inappropriate content was an accident or not, go talk to a trusted adult about it. Talking about it can ease a huge weight. This trusted adult can help set up precautions to protect you from accidentally seeing it again or can help you take steps to stop looking at it.

QUESTION...

What should I do if an online "friend" wants to meet?

ANSWER:

Never meet up with someone you have met online. Children and teens are targeted for all types of crimes this way. Tell a parent if someone wants to meet you in person. If your parent feels it is okay, go with your parent and meet in a public place.

QUESTION...

Are there pictures that should not be posted online or sent by a phone?

ANSWER:

Absolutely. It is vital to remember that things you send through text or post online can never be erased permanently. They could be seen, even years down the road, by prospective employers, colleges you apply to, neighbors, and even your future children. Before sending any photo, ask yourself, "Would I send this to my mother?" If the answer is no, don't send or post it.

QUESTION...

What should I do if someone sends me an inappropriate photo?

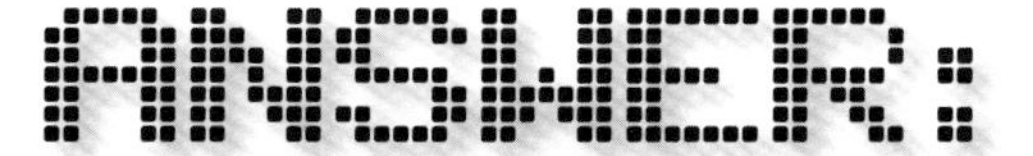

Did you know that if you have an inappropriate photo of someone under the age of 18, it could be considered possession of child pornography, even if the person is your own age or older? This is a serious crime that can lead to prosecution and a requirement to be registered as a sex offender. If anyone sends you an inappropriate photo, tell an adult immediately, and block the person sending the photo(s).

QUESTION...

How can I make sure I am a good online citizen?

ANSWER:

1. Remember the Golden Rule. Arguing, insulting, and bullying can have real-life consequences; choose to be respectful.
2. Don't interact with mean people.
3. Don't copy and paste other people's images, texts, or videos and claim them as your own.

4. Don't post something that could be harmful to yourself or others now or down the road.

5. Be extremely careful with your personal information, such as your phone number, address, and social security number.

6. Don't believe everything you read online.

QUESTION...

How can I best use the internet for fun and learning?

Content for fun and learning online is endless. From studying a new language to connecting with friends and family to learning how to build a car engine, the opportunities go on and on. If you know about the internet's pitfalls, you are less likely to fall into them, thereby creating a more positive and fun learning experience.

QUESTION...

What is cyberstalking?

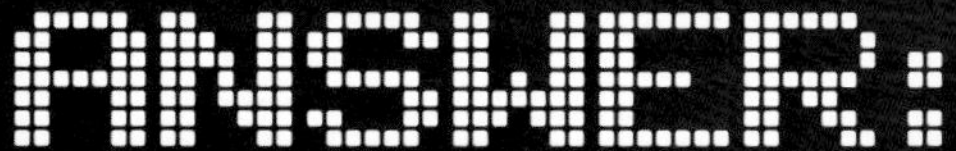

If you are being monitored online or receiving unwelcome messages from either a stranger or even someone you know, that could be called cyberstalking. It may be motivated by anger, revenge, control, lust, or a number of other reasons. If you feel you are being cyberstalked, let a parent or trusted adult know. Some cases may even need to be reported to the police.

QUESTION...

How can my webcam be used to harm me?

ANSWER:

Some cybercriminals are able to hack into the camera on your computer and spy on you. They are usually watching for content they can use to blackmail you. Pay attention to the light next to your webcam. If it comes on at random times or without any action from you, there may be malware on your computer turning it on. Run a scan to remove malware. Ask an adult to help you do this. You can also get a webcam cover for added security.

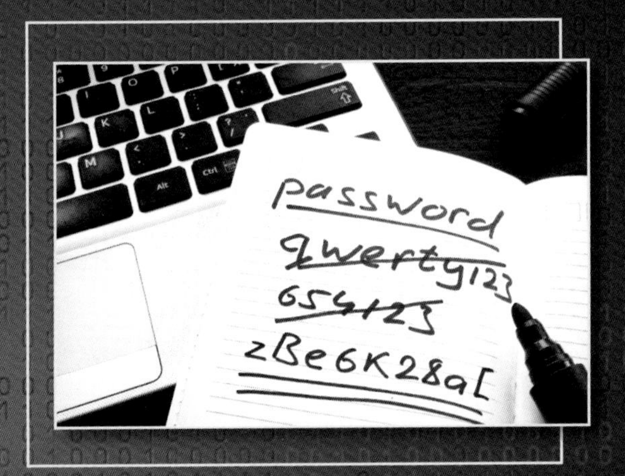

QUESTION...

Why is it important to keep my passwords secret?

ANSWER:

Cybercriminals pretending to be your bank or your internet service provider may ask you for your password. Never give passwords to anyone besides your parents. Banks and legitimate companies will never call or email you asking for passwords. If in doubt, go straight to the website and log in. Don't click on email links and log in. Even friends may not be responsible with your passwords. They are best kept between you and your parents.

QUESTION...

What is hacking?

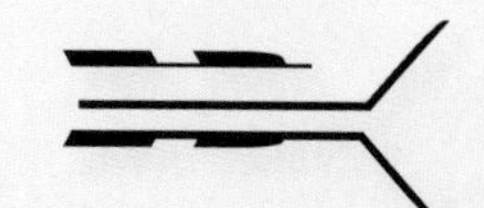

This is when cybercriminals gain unauthorized or illegal access to computers or computer systems by figuring out codes and passwords or bypassing them.

QUESTION...

What are the risks to online shopping?

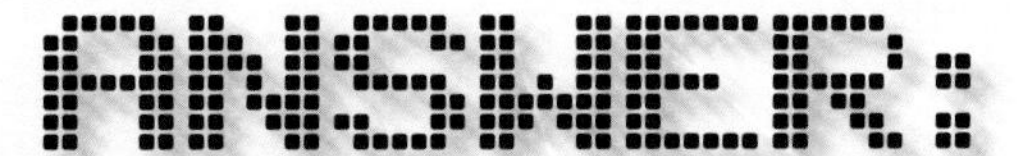

Online shopping can lead to viruses, scams, or the stealing of your personal information. It is important to make sure that any website where you shop has the padlock icon next to the web address, signaling that the website is secure. Also, don't give more information than needed. For instance, no purchase should ever require your social security number. Before the age of 18, you should make purchases only with parental approval. If in doubt, do a search of the website with the web address followed by the word "security" and see what information you get.

QUESTION...

What should I know about cookies?

ANSWER:

We're not talking about the warm kind with melty chocolate chips... here we are talking about bits of information saved on your browser or to your computer. Generally, these bits of information help you have a smoother, more pleasant browsing experience because you don't have to reenter information each time you visit a website. However, some cookies are used to track your browsing history in order to send you custom advertising. Many people feel this is a serious breach of privacy. There are ways to clear cookies from your browser and to change settings to prohibit cookies on your computer. Ask an adult to help you do this.

QUESTION...

Why are rules about family internet use important?

ANSWER:

One in five children receive sexual solicitations online. There are always cyberbullies and predators online targeting children. This is not meant to scare you, but the more you know, the better you can protect yourself. Parents set rules to keep you safe physically and emotionally. If your parents have not set rules, it is a good idea to talk to them about the risks and set some rules together.

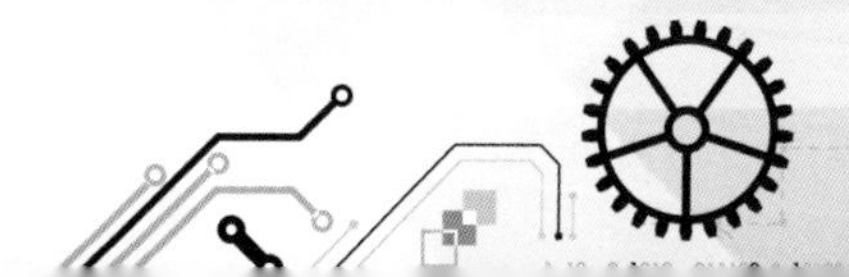

QUESTION...

What if I am using the internet at a friend's house? Do I still have to follow the rules for my house?

Yes. Always follow the rules you and your parents have set. Your friend may also have house rules about computer/internet use, and you need to honor and respect those rules as well.

QUESTION...

Why is it important to keep open and honest communication with my parents about my online experiences?

ANSWER:

Open communication keeps you safe. The biggest reason children fall victim to online predators is they are afraid to tell someone what is going on. Predators are very good at manipulating children. After becoming "friends" and offering a child money, popularity, gaming advantages, and so on, they will later threaten the child or make him or her feel ashamed. But remember, these crimes ARE NOT YOUR FAULT. Your parents will help protect you and ensure positive online experiences.

QUESTION...

What is spam, and how can I avoid spreading it?

ANSWER:

Spam is any unwanted, annoying digital communication that is sent out in bulk. It usually comes through email, but it can also be received through social media, texts, or phone calls. Be careful of chain emails that ask you to forward things to friends. Often spammers are just collecting masses of emails to use to send out more annoying spam or malware.

QUESTION...

What can I do if someone I know is engaged in harmful online activities?

Talk with them about it, and also consult with a parent or other trusted adult.

QUESTION...

If I am on a website, can I give out my name and phone number to enter a contest?

ANSWER:

Websites should NEVER ask children for identifying information. Always ask a parent for permission before you give ANY personal information. If the website is asking for this information, chances are it is a malicious website.

QUESTION...

What should I do if I'm online and my internet provider asks for my password because the company needs it to fix my account? Should I give it to them?

NO. Legitimate companies will never ask you to reveal a password.

QUESTION...

What should I do if I see a negative post about me online?

ANSWER:

This is usually called cyberbullying. Here are some steps you can take to resolve the problem:

- Tell a parent what is going on.
- After printing or taking a screenshot as evidence of the comment or post, delete it (if possible).

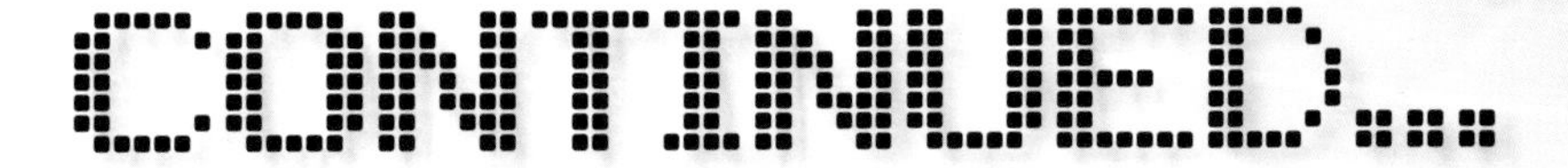

- Don't engage the bully by commenting.
- If a post or photo is distressing, untag it and flag it for removal.
- Block or unfriend the person who is leaving negative posts.

If these steps don't end the problem, you can talk to a school or church counselor, with your parents' permission. In extreme cases, the police may become involved.

QUESTION...

Can I trust everything I read online to be true?

ANSWER:

The sky is falling! The sky is falling! There are a LOT of exaggerations, false claims, and outright lies online. It is also good to recognize bias in things we read. Bias means a tendency to have an unfair perspective or to write something with prejudice. As we stay close to God, He will help us recognize truth from falsehoods and know good from bad. Like Solomon who prayed, "Give therefore thy servant an understanding heart . . . that I may discern between good and bad" (1 Kings 3:9), you can also pray for help to discern truth from error and good from evil. Parents can often see red flags that you don't, simply because they are more experienced.

CYBER SAFETY PLEDGE

1. I will be mindful of the time I spend online, on a phone, or with other technology. I will make sure my use of technology doesn't interfere with real human interaction, schoolwork, and sleep.

2. I will neither tolerate bullying nor be a bully. I will respect myself and others.

3. I will be a kind online citizen. I will support friends and others who may need help with difficult situations.

4. I will strive to create an atmosphere of tolerance and respect at home, at church, and with my peers.

5. I will help my parents and other family members when they need my assistance with the use of technology.

6. I will practice smart internet security and guard my passwords.

7. I will be respectful of the digital property and space of others. I will never break into other people's accounts or use their content without permission.

8. I will be conscientious of my use of forwarding and copy and paste. If I use other people's content, such as writing or photos, I will be sure to give credit by quoting them or including a link to their content.

9. I will respect the privacy of others, remembering kindness and courtesy if I post comments about or photos of them.

10. I will never send or post photos or other material that could embarrass me, threaten my privacy or safety, or get me or others into trouble.

1.0NSP239-148190